IMAGES
of America

SEBASTOPOL'S GRAVENSTEIN APPLE INDUSTRY

On the Cover: The ticket office of the first Sebastopol Gravenstein Apple Show at South Main Street drew crowds young and old. The huge tent behind the office facade held dozens of apple displays from local growers, packinghouses, and canneries promoting their products. (WSCHS.)

IMAGES
of America

SEBASTOPOL'S GRAVENSTEIN APPLE INDUSTRY

Western Sonoma County Historical Society

ARCADIA
PUBLISHING

ISBN 978-0-7385-8173-6

Published by Arcadia Publishing
Charleston, South Carolina

Printed in the United States of America

Library of Congress Control Number: 2010932075

For all general information, please contact Arcadia Publishing:
Telephone 843-853-2070
Fax 843-853-0044
E-mail sales@arcadiapublishing.com
For customer service and orders:
Toll-Free 1-888-313-2665

Visit us on the Internet at www.arcadiapublishing.com

This book is dedicated to the Sebastopol apple community who lived this history.

Contents

ACKNOWLEDGMENTS

We wish to thank the following people without whose help this book would not have been possible. Research and text: Frank Baumgardner, Evelyn McClure, Stacy Ruppert, and Rae Swanson. Technical support: John Fore, Sally Morrison, Tom Perry, Betty Short, and Steve Stedman.

Photographs were provided from the collections of the people or organizations listed below. In most cases an abbreviation follows the contributor's name; these are used throughout the book to denote those images. They are as follows: Al and Cecilia Mello (ACM); Bill Silva (BS); Carol Capria (CC); Coon family (CF); Don Roberts Collection (DRC); Donna Hallberg Marovich (DHM); Don Meacham-Sonoma County Library (DM-SCL); Dorothy Watson Dorward (DWD); Evelyn McClure (EM); Dorothy Bridgeford (DB); Edward Trigeiro (ET); Harold Narron (HN); Hazel Singmaster Jens (HSJ); Hyla Tough McDonell (HTM); Joan Voit Chilton (JVC); J. P. Munro-Fraser, *History of Sonoma County, 1880* (MF); Joe Moir photograph, courtesy of John and Kristina Schmale (JM); Joy and Gene Shreyer (JGS); Kashiwagi family (KF); Lee and Shirley Walker (LSW); Lillian San Filippo (LSF); Lorna Skinner (LSC.); Manzana-Norton and Sandborn (MZ); Merle Avila Collection (MA); Merlyn Winton (MW); Pat Marshall (PM); Phyllis Van Winkle (PVW); Robert Schilling (RS); Rose and Tetsuro Fujii (RTF); Rosebrook family-Wendy L. Deakins (RF); Doris Rowe Collection (RC); Russian River Slow Food Presidium, Paula Shatkin, Coordinator (SFRR); Sebastopol Chamber of Commerce (SCC); Sonoma County Library (SCL); Sonoma County Recorders Office (SCRO); San Francisco Examiner-Lenore Carrion used with permission of UC Regents-Bancroft Library, Berkeley, California (SFEX); Sonoma West Publishers (SWP); Tia Polley; The Winkler Family Association, 1988 (WFA); Thomas D. Barlow (TDB); Tom Perry (TP); Western Sonoma County Historical Society Archives (WSCHS); William S. Borba Collection (WSB); and Zimpher family (ZF).

Introduction

The California Gold Rush of 1849 brought many people seeking easy riches. When that proved to be false effort for most would-be miners, they turned to what they knew best, farming. Early white settlers found their livelihood, if not riches, in Sonoma county soil.

Joel Walker and his family had been introduced to the promise of the great abundance that California held by his brother Joseph Walker, who had led exploration parties throughout the West. The Walkers' son John and his friend Joseph Millar, a physician by training, opened a trading post in 1849, south of today's Sebastopol. Joseph Morris arrived in the area in 1853 from Ohio. He worked for Millar and Walker and then claimed 120 acres of government land situated between two already established ranchos. And so, in 1855, Morris founded Pine Grove, Sebastopol's original name, between the vast acreage of Jasper O'Farrell's Canada de Jonive rancho and the Llano de Santa Rosa rancho of Joaquin Carrillo.

Pine Grove became Sebastopol as a result of the notorious siege of Sebastopol during the Crimean War. The Russian army held the Sebastopol peninsula from invasion by British and French forces to the point of starvation. In Sonoma County, Jeff Stevens and Pete Hibbs got into a tussle in Pine Grove. Hibbs took refuge in John Dougherty's store, and the storekeeper refused to let Stevens in to continue the fight on his premises. Stevens kept watch outside until he lost interest in the fight. The gathered crowd proclaimed this was Hibbs's Sebastopol. Somewhat later, since there were other Pine Groves in California, when an official post office was to be established, the residents opted for the name Sebastopol for this growing community.

By 1876, with Sebastopol entering into its 21st year of existence, it boasted 40 homes, three general and grocery stores, one hotel, one livery stable, three saloons, a blacksmith, one butcher shop, a shoemaker, a lumberyard, two churches, a school, two teachers and 45 pupils, two physicians, and a population of 250. But city lot sizes were large; sometimes properties were many acres so that orchards and gardens lent a very agricultural air to the small settlement.

Sebastopol is not quite ready to look back on the Gravenstein apple with nostalgia. Its orchards are struggling, but alive, if not well in the area. Sonoma County agriculture began in the early 1800s, when the Russians at Fort Ross planted an apple orchard and other crops. Their trees were said to be still heavily bearing apples into the 1880s. Mitchell Gilliam and his son-in-law Isaac Sullivan arrived in Green Valley, today's Graton area, in the early 1850s. They bought apple tree stock in Petaluma and planted an orchard. They traveled by horse and wagon, selling their fruit door-to-door and in towns. The Sebastopol area quickly gained a reputation for its abundance of apple orchards and other agricultural crops. The area was termed "The Gold Ridge" by the *Analy Standard* newspaper owner, Mary B. Williams. The usage of that term continues today.

In the mid-1800s, the county began producing crops of grain to feed cattle, sheep, and horses. The mission fathers planted wine grapes followed by other pioneer grape growers such as Agoston Haraszthy. California grape growers did not need encouragement to grow grapes once they discovered varieties that did well in the state; unlike the Virginia House of Burgesses that, in 1619, passed a law requiring every male colonist to plant 20 vines. For each dead or non-fruit-bearing vine, the fine was a barrel of corn. The grape varieties they planted did not flourish, and the House of Burgesses came into possession of much corn. Sebastopol had several wineries in operation in the 1890s. Emile Schirmer's Belle View Winery, John Bonnardel's Winery, and the C. F. Julliard operation processed the local grape harvest. The Julliard winery shipped several carloads of wine a day for a month to make room for the new vintage. The Women's Christian Temperance Union had an uphill battle with all the local alcoholic beverages available.

Sonoma County's soil was most productive for many crops. Potatoes were grown in great quantity, a crop familiar to Irish settlers. Dairy farms were established by Swiss Italians. Then came the hops boom until it was wiped out by Prohibition. Peaches and cherries were grown extensively, then succumbed to blight. Berries were planted between rows of newly planted orchards to produce income until the apple trees matured. W. J. Hunt planted 10 acres of blackberries, which he sold fresh and planned to make 2,000 gallons of wine for medicinal purposes. In 1900, farmers became more interested in berries, and a dozen berry fields were planted. One extensive strawberry grower was A. E. Scammon, the son of a whaling captain who settled in Sebastopol along the Laguna watershed for several years. In 1905, Walter Armstrong, the son of Col. James Armstrong (for whom Armstrong Redwoods State Reserve in Guerneville is named), was growing 7 acres of asparagus. His ranch, known as Fruitland Farm, was located 2 miles west of Sebastopol. He planned to expand his asparagus plantings to 40 acres. He also grew rhubarb extensively, had 4,000 apple trees, and a large vegetable garden.

Apples and other fruit were being produced in Green Valley in 1870. There was enough volume (20,000 boxes) for growers to meet with the San Francisco and North Pacific Railroad owners to explore new ways to ship fruit to San Francisco. Previously, farmers hauled fruit to Petaluma by horse-drawn wagon. It was then loaded onto ships bound for the city. The San Francisco and North Pacific line only came to Santa Rosa from Petaluma. Its branch line into Sebastopol was not built for another 20 years, but by 1876, fruit growing, stock raising, and farming were reported to be the principal interests supporting the town. Fruit dryers were being constructed too and rebuilt after the predictable fires. A cannery was operating seasonally in Sebastopol in 1887, employing 15 workers one day a week. They canned pears, peaches, blackberries, and apricots. Within 10 years, the cannery employed up to 325 people in peak season. Frank Harrison's former blacksmith shop was converted to a packing shed, and fruit was sent out from there to Tacoma and Seattle, bringing in $9 a ton. Farmers' fruit meant not only food for their families but cash income. Sampling the farmers' wares from the field uninvited was not encouraged. In 1897, four boys on a camping trip helped themselves to peaches in James McChristian's west Sebastopol orchard. They were waylaid in Bodega, hauled back to justice court in Sebastopol, and fined $2 each. Today boxes of apples could mysteriously appear at your doorstep along with the ubiquitous zucchini whether they are welcome or not.

One

From Italy to Sebastopol

A warm piece of apple pie is the quintessential American dessert. One of the best apples used for baking, especially in an apple pie, is the Gravenstein apple. Lucky for Sonoma County residents, this unique apple can be found in abundance. The Gravenstein is well known for its yellow-green skin with red stripes, along with having a tart but sweet flavor. Although Sonoma County is now known for the grapes and premier wine sold here, the Gravenstein apple had its own place in the diverse agricultural history of the area. Sebastopol's temperate climate and fertile soil allowed Gravenstein apple trees to thrive. In turn, the town would grow commercially due to the Gravenstein and its unique characteristics. It is the first type of apple to ripen in either late July or early August and keeps its flavor whether cooked, made into applesauce, canned, or dried. Its ability to keep while being shipped across the world also helped to increase the popularity of the fruit. While Sebastopol produced other apples, the Gravenstein was by far the most successful agricultural product for the town. Sebastopol would become well known not just in California but in Europe as well, due to the Gravenstein. The Gravenstein would become so entrenched in Sebastopol life, that celebrations were held in its honor.

The origins of the Gravenstein in Sebastopol, and throughout the world, have been debated over the years. Some argue the Gravenstein originated in Italy, while others believe it came from northern Germany or Denmark. A popular theory combines all three locations, but with the first tree growing in Italy. In one story, Prince Carl of Denmark and his friend Count Ahlefeldt visited Italy in the late 17th century. While in Italy, they ate an unusually delicious apple. Count Ahlefeldt took some seeds and cuttings from the tree back to his estate, Castle Graasten, in Schleswig, Denmark. The Duke of Agustenborg would later buy the castle and rename the property with a more German-sounding name to fit his heritage—Castle Gravenstein. In the late 18th century, a nurseryman named Livoni in Sonderburg, Denmark, began working with Gravenstein cuttings to produce an even better tasting apple. The Gravenstein would be planted throughout Europe due to its unique flavor. This 1940s Sebastopol Apple Growers Union poster advertised the "fresh from the orchards of California" Gravensteins. (WSCHS.)

Most historians agree that Russians living at Fort Ross brought the Gravenstein apple trees to Sonoma County in the early 19th century. The trees found growing on the hillside near the fort most likely came from seedlings the Russians had gotten from Europe and descended from Livoni's plants. In the early part of the 20th century, Nathaniel Griffith, said to be the "Grandfather of the Gravenstein," and Luther Burbank, botanist and horticulturist, found a Gravenstein tree that was over 100 years old, growing at Fort Ross. The two would trace the roots of the Fort Ross tree back to Germany, seeming to confirm the theory that the Gravenstein the Russians had did originate from Germany. Old Fort Ross is seen in a 1990 photograph above, and as depicted, appropriately, made of Gravenstein apples at the 1939 Gravenstein Apple Show. (Above, EM; below, SCL.)

It is also possible that the Gravenstein came from a local nursery started by Henderson Luelling. In 1854, nurseryman Luelling opened a nursery in Oakland, which sold thousands of trees, including the Gravenstein. Sebastopol growers may have purchased trees from the nursery, but there is a large gap between when the trees might have been purchased and when the first commercial apple orchards began in Sebastopol. Another possibility was that while the Russians were bringing Gravenstein seedlings to the west, emigrants from Denmark were bringing seedlings to the East Coast. When those emigrants started moving west in the 1870s, they brought seeds and cuttings with them. Regardless of its start in Sebastopol, the Gravenstein would create an entire industry of farmers, apple pickers (as seen in this 1920 photograph harvesting Gravensteins in a Sebastopol orchard), packers, factories, and processing plants. (Both, WSCHS.)

In 1883, Nathaniel Griffith planted what would become the first commercial Gravenstein apple orchard in Sonoma County on his Laguna Road property. With Luther Burbank's help, Griffith increased the yield of his Gravensteins and would eventually become one of the largest Gravenstein growers in the county. Griffith believed the Gravenstein would gain in popularity, and he set out to prove the economic value of the fruit. According to his friend Burbank, "if the Gravenstein could be had throughout the year no other apple need be grown." By 1890, the first Gravenstein crops were being brought to apple brokers in Sebastopol and later crated and taken to San Francisco by railroad cars. Griffith is seen next to his beloved Gravenstein orchard, and his home and property are seen on Laguna Road, both around 1930. (Above, JGS; below, CF.)

As Griffith predicted, the Gravenstein apple would become an incredibly popular agricultural crop in Sonoma County. The Gravenstein thrived in the climate and soil unique to the Sebastopol area and did not need much irrigation. This superb land would be nicknamed "Gold Ridge," partly due to the prosperity of the Gravenstein. In 1890, orchard land sold for $20 per acre. By 1923, the price of an acre had increased to $2,000, and in 1979, one piece of orchard land was advertised at $10,000 an acre. Aerial photographs show two views of the Gold Ridge area. This c. 1940 photograph (above) shows Main Street Sebastopol from upper left to mid right with orchards visible near town. The second photograph, taken in 1961, is looking west from Sebastopol along Bodega Highway with a myriad of apple orchards that dot the landscape. (Above, DRC; below, SCRO.)

Since Gravensteins growing on the East Coast and Canada ripen two to four weeks after Sebastopol's, local growers were in the unique position to be able to ask for any price for their apples at market. About 100,000 tons, or 500 carloads, of Gravenstein apples were produced and shipped from Sebastopol to be sold at market in 1911. By 1919, fresh fruit shipments consisted of 525 carloads of Gravensteins and about 18,000 tons of dried fruit, cider, and vinegar made from apples. A box of freshly packed Gravensteins would sell for between $2.25 to $2.75. Workers in a local packinghouse (above) prepared apple boxes for shipment around 1900. Women were a large component of the apple industry workforce. At right, they are depicted in a 1914 Gravenstein Apple Show exhibit working "in the shade of the Old Apple Tree." (Above, WSCHS; at right, MA.)

In this 1920 apple exhibit, Miriam Hotle Burdo points to the many locations Sebastopol's Gravenstein apples and canned applesauce were shipped. The Gravenstein's ability to ship around the world and still maintain its quality and flavor was demonstrated in the South American and South African markets. It would take at least five weeks before a Gravenstein shipment would reach its destination (by train and ship), but the fruit would look as fresh as when it was packed in Sebastopol. The Gravenstein apple growers were able to profit even more during World War I, with the high demand of dried fruit in war-torn Europe. This 1910 first-place winner apple show exhibit of Northwest Pacific locomotive engine no. 22 is symbolic of the role the railways played in the Gravenstein's success. (Above, HTM; below, MA.)

As the Gravenstein rose in popularity throughout the world and became an integral part of life and economic vitality in Sebastopol, a celebration would be organized in its honor. In 1910, the Sebastopol Apple Growers Union started the Sebastopol Gravenstein Apple Show. The union not only wanted to celebrate the fruit, but also increase the local market for Gravensteins by promoting apple products and growers. A huge tent was erected on Main Street across from the Petaluma and Santa Rosa Railroad depot, now the West County Museum. At this first 1910 apple show, a large group of uniformed boys from the San Francisco Children's Aid Society were lined up waiting to enter. These boys, and also girls, from San Francisco came to Sebastopol and Graton each summer to help with the berry harvest. (Above, WSCHS; below, ET.)

The early Gravenstein show organizers named their group the Gravenstein Apple Show Association (GASA). Each year, the GASA were photographed at the opening of the show near one of the exhibits as this group above is in 1912. The organizers usually included local merchants, fruit brokers, realtors, bankers, a mayor or other local official, and of course apple growers. Another group photograph, this time of 1915 show attendees lined up to enter the show all dressed in their Sunday best is below. In the background, a portion of one of the fair tents is visible, as well as an archway made of Gravensteins. The apple show was a big event in the life of Sebastopol residents and visitors, alike. (Above, ACM; below, HSJ.)

From the start, the Gravenstein Apple Show featured apple sculptures made from all parts of the apple and depicted everything from local landmarks to nursery rhymes, historical events, and animals. If it could be created from an apple, display makers designed it. One of the 1910 apple displays depicted the name of Sebastopol in a clever way. Instead of spelling out the name, Sebastopol was depicted phonetically from a "C" to a bass fish to a pool at the very end of the display. At the beginning of the display, there is a watch made out of apples. The entire message reads "Watch Sebastopol." The 1911 apple show featured a miniature Sebastopol apple farm, of course made from apples, with the Gold Ridge hills in the background. (Above, WSCHS; below, MA.)

Although the Gravenstein Apple Fair was held in downtown Sebastopol, many of the other surrounding towns comprising Sonoma County also participated in the event. Throughout the fair's history, examples of town pride can be seen in the displays they created showing off their own iconic buildings and landscapes. In the 1910 show, a group from Forestville created a replica of their schoolhouse made entirely of apples from the foundation to the very top of the bell tower. We can only marvel at the ingenuity and creativity—not to mention painstaking work that went into these exhibits. Creating the schoolhouse was almost prophetic, as the Gravenstein Apple Fair would be held 66 years later at the Forestville Youth Park, which was not too far from the school. This first fair was very festive and patriotic, as seen with the American flags hanging from the tent rafters. (SCL.)

The apple show featured apple sculptures celebrating local landmarks such as the First National Bank of Sebastopol, the Sebastopol Gristmill (which was built in 1852), and Analy Union High School. Gristmills were usually waterwheel operated and they were utilized to grind grain into flour—a very useful function in pioneer times. Analy Union High School was only about three years old when it was reproduced in apples at the 1911 show. There were five seniors in the first graduating class in 1910. Although the apple displays were extraordinary the first year, each year they became progressively more elaborate, while still celebrating local life and historical events that touched the small town. (Above, MA; below, WSCHS.)

In 1911, the Gravenstein Apple Show had an apple display of the steamship *Gold*. Through the Petaluma River, the *Gold* transported apples, fruits, and other produce and goods from Sonoma County to San Francisco to be sold at market. True to all the displays from these early years of the apple show, the steamship replica included amazing detail; the *Gold* sat in a pool of water and included a rotating paddle wheel. The exhibit, not surprisingly, won first place. In 1915, another part of local history was exhibited with the Santa Rosa District exhibit, entered by the chamber of commerce, and displayed Luther Burbank's Santa Rosa home. (Both, MA.)

Whole, dried, and sliced apples were not only used to recreate landmarks, but also to depict how apples were packed, weighed, and shipped. An apple scale was created for the 1914 fair using apples and even showed apples being weighed. The 1911 display at right had a boy climbing an orchard ladder made, appropriately, of apples. Besides the apple displays, other events of the Gravenstein Apple Fair included horse races down Main Street, live music, and apple pie contests. With the many apple growers' displays, fair attendees would see 4,000 boxes of apples under the huge tent. At the 1910 fair, there was even a display that included apples from the Russian orchard at Fort Ross. Colorful banners and American flags were hung near the fair's tent and throughout downtown Sebastopol. Nearly 25,000 people attended the first show, arriving in horse-drawn carriages. (Above, WSCHS; at right, MA.)

In 1912, a display of the ocean liner *Titanic* sinking into a pool of water was created from whole, dried, and sliced apples. The iceberg that sunk the *Titanic* was created using sliced apples and a placard with the name W. S. Liddle on top. Liddle was busy cycling around the world during the time the *Titanic* sunk, and it was a local joke that the ship had sunk when the two collided along their separate journeys. Liddle had decided to undertake cycling around the world in order to be the first Britisher to accomplish the feat, to keep and have his diary published, and also to make a name for himself. Liddle would eventually settle in Sebastopol. (Both, MA.)

Each year, apple display designers became more creative in the subjects they decided to create and how they would make the displays come to life. In 1913, many local landmarks were once again depicted such as the town of Sonoma and the June 14, 1848, Bear Flag revolt. The Graton community created a clock tower complete with a swinging pendulum that won first prize that year. Other displays included the Stony Point district's Chute the Chutes and a gigantic reproduction of a phonograph. More international landmarks included a replica of the Parthenon and a recreation of the red-and-white-striped Eddystone Lighthouse located in Plymouth, England. (At right, HSJ; below, WSCHS.)

Most people know of the infamous dirigible *Hindenburg*, which exploded in 1937. However, before the *Hindenburg*, dirigibles had been used as passenger and cargo ships from around 1910 until 1937. During World War I, nonrigid dirigibles, or blimps, were used to observe the enemy because of their ability to hover better than airplanes of the same time period. With the popularity of the dirigible, it was no wonder this helium-filled airship was part of the 1913 Gravenstein Apple Show. A life-sized dirigible was created entirely from apples and hung from the exhibit tent. Although it must have been quite heavy, this amazing sculpture seemed to float above the rest of the exhibits. (Both, MA.)

At the 1915 Gravenstein show, one group created the Liberty Bell entirely out of apples, complete with its famous crack. Another display was of a carousel. To create the feel of a carousel, apples were draped to create a circus tent top, and the entire platform and poles were created from apples. The only objects not created out of apples in the display were the platform horses. The fair would see certain themes repeated throughout the years, including common rides at fairs such as a Ferris wheel and a Flying Dutchman ride in 1912. Another theme that would develop throughout the history of the Gravenstein Apple Fair would be scenes from nursery rhymes. (Both, PVW.)

A popular nursery rhyme, "Little Jack Horner," was depicted at various fairs. Nursery rhymes were often based on historical events or people. The Jack Horner rhyme started in the 16th century and is said to be about Thomas Horner, who was the steward to the Bishop of Glastonbury. Horner was sent by the bishop to give King Henry VIII twelve title deeds to various estates hidden in a Christmas pie. On his way to the king, Horner reportedly stole the deed to Mells Manor, the "plum" of the 12 estates. Mells, located in France, would become the property of the Horner family, where relatives are still living today. The rhyme goes: "Little Jack Horner / Sat in the corner / Eating a Christmas pie / He put in his thumb, and pulled out a plum / And said, 'What a good boy am I.' " (Both, ET.)

"Peter, Peter, Pumpkin Eater / Had a wife and couldn't keep her / Put her in a pumpkin shell / And there he kept her very well," goes the rhyme. In the 1930s, "Peter, Peter, Pumpkin Eater" was created and displayed by the town of Forestville. The nursery rhymes told today are often rooted in British history and were in fact a way to spread gossip and rumors about English royalty. As the rhymes were retold through the centuries, the true history of many of the stories was lost or changed, as often happens with rumors. One theory about the story of Peter, Peter, Pumpkin Eater is that a poor man had an unfaithful wife. To keep her from cheating on him, he put her in a chastity belt (the pumpkin shell in the rhyme). With the chastity belt, Peter was able to keep her from finding other suitors. (WSB.)

"Humpty Dumpty sat on a wall / Humpty Dumpty had a great fall / All the King's horses, and all the King's men / Couldn't put Humpty together again," goes the nursery rhyme. The Sebastopol Chamber of Commerce recreated Lewis Carroll's Humpty Dumpty from his *Through the Looking Glass*. The wall Humpty is sitting on is a wall of Gravenstein apples. Unlike other nursery rhymes, "Humpty Dumpty" was actually about a large cannon used in the English Civil War in the Siege of Colchester in 1648 between the Royalists and Parliamentarians. The Parliamentarians damaged the wall underneath Humpty Dumpty causing it to fall to the ground. The Royalists tried to put the cannon on a remaining portion of the wall, but it was too heavy. Without this important piece of weaponry, the Parliamentarians were able to win the battle that lasted 11 weeks and take over Colchester. (ET.)

"Hey diddle diddle / the cat and the fiddle / the cow jumped over the moon / The little dog laughed to see such fun / and the dish ran away with the spoon," goes the rhyme. Portrayed in an apple display by Sonoma Valley, the history of this rhyme is controversial. The most popular theory is that the rhyme is about Elizabeth I, who was nicknamed "The Cat" because of the way she played her cabinet members. The "little dog" is supposedly Robert Dudley, who was the Earl of Leicester, and who Elizabeth nicknamed her lapdog. The cow and moon are supposedly nicknames of members of her court. The dish and spoon are said to represent two of Elizabeth's servants (a serving lady and royal taster) who fell in love and eloped and were later imprisoned in the Tower of London by Elizabeth's order. (ET.)

Unlike some of the other rhymes, "Seesaw Marjorie Daw" was probably just a funny rhyme made up for children to sing while playing on a seesaw. However, the last three lines probably reflect the use of child labor in workhouses, especially for those children who did not have anywhere else to live and were forced to work for 1¢ a day. The rhyme goes: "Seesaw Marjorie Daw / Johnny shall have a new master / He shall earn but a penny a day / Because he can't work any faster." In the background of the Seesaw Marjorie Daw display, created by Geyserville, another display by the Sebastopol Apple Growers Union can be seen. It is a depiction of the "Old King Cole" nursery rhyme, which is based on British history in the third century when the empire was ruled by Celtic kings. (Both, ET.)

"Cock a doodle doo / My dame has lost her shoe / My master's lost his fiddling stick / And knows not what to do." These four lines of "Cock-A-Doodle-Doo" first appeared around 1765, with the other lines of the rhyme being added much later. In one version of the rhyme's history, it is said it tells the story of a young girl who witnessed the murder of her young brother. In order to ensure she would not talk about what happened, the murderer cut out her tongue. Later when the girl was being teased by a group of children, she was able to say the words back to them with the encouragement of a friend. Sebastopol historian William S. Borba was the photographer of this c. 1930 display. Borba's lifelong passion was documenting Sebastopol's and Sonoma County's history. (ET.)

"Mary, Mary, quite contrary / How does your garden grow? / With silver bells and cockleshells / And pretty maids all in a row," goes this rhyme. The Santa Rosa Chamber of Commerce created another rhyme steeped in British history in this *c.* 1930 display. "Mary, Mar,y Quite Contrary" was actually about Queen Mary Tudor, also known as "Bloody Mary." During her reign, Queen Mary had anyone who practiced Protestantism executed since she was a Catholic. The garden in the rhyme actually refers to the multitude of graveyards of those who continued to practice the Protestant religion and were executed because of it. The silver bells and cockleshells represent instruments of torture used during the time. The maids in the rhyme were actually guillotines, which had been known as the maidens. On a happier note, one of the displays in the left background is the Women's Auxiliary Pie Booth—apple pie, no doubt. (ET.)

"Ding-dong bell / Pussy's in the well / Who put her in? / Little Johnny Flynn / Who pulled her out? / Little Tommy Stout / What a naughty boy was that / Try to drown poor Pussycat / Who ne'er did any harm / But killed all the mice / In the Farmer's barn!" Unlike some of the previous rhymes, "Ding Dong Bell" actually has nothing to do with British royalty or hidden meanings of cruelty. The rhyme dates back to the 16th century and is actually a rhyme to teach morality to young children. The message is that it is wrong to hurt an innocent animal. However, when the rhyme was first written, it ended with the cat being left to drown. The line was later changed to "Little Tommy Stout." (ET.)

Sebastopol's Post No. 39 of the American Legion sponsored this Gateway to Mother Goose Land exhibit in the 1930s. Apple rings, whole apples, and apples in crates helped create the entrance to the magical land of Mother Goose. Mother Goose has as interesting history as the nursery rhymes she is often identified with, although much less bloody. The name Mother Goose was first created as part of a poem published in the mid-1600s. It would not be until 1780 that she first appeared with nursery rhymes in Britain. In 1786, she would be a part of an American edition of rhymes. Mother Goose's popularity slowly grew through the centuries, as she became the popular keeper of childhood nursery rhymes. (ET.)

Various businesses exhibited their products with booths and displays at the fair even though they were not a part of the apple industry. Pacific Gas and Electric created a special electrical display at the 1912 fair. A number of florists in Sebastopol and Sonoma County also had booths and displays at the fair depicting their products. The Fred Grohe florist shop was a repeat exhibitor at many apple shows throughout the years. These interesting displays not only showed fair visitors the different goods Sebastopol and Sonoma County had to offer but also were unique advertisements for local businesses and their products. (Both, WSCHS.)

This 1910 apple exhibit is of the California State Capitol building in Sacramento, although missing its pediment and two "drums" that support the dome. The state capitol building was begun with a ground-breaking ceremony in 1860 with a budget of $100,000. Fourteen years later in 1874, it was completed with the finished price tag of $2.5 million. This apple sculpture is quite detailed in its application of sliced apples, even covering the steps up to the capitol. Note the handwritten "Lost" sign placed against the building. On the sign, the author wrote: "Lost! On Healdsburg Road between Sebastopol and the residence of T. J. Compton, lady's handbag containing railroad ticket of Miss Belle Vorfel to St. Louis. Also glasses, gloves, papers and small change. Finder please return to John A. Williams. At Sandborn's Ranch." Hopefully Belle Vorfel found her way home to St. Louis after visiting the apple show. (WSCHS.)

The 1910 show program quoted Luther Burbank who best described the Gravenstein as follows: "The Gravenstein apple has, above all others, proved to be the money winner in Sonoma County. It is a healthy, vigorous tree. It always bears a good crop, never overbearing, as many varieties do; is of the best quality of all known apples." The Sherwin Williams Arsenic of Lead Company awarded Ida Hunt, wife of apple pioneer William J. Hunt, a special cup prize for her apple display in 1910. The silver cup is seen to the right of the young man above. In 1911, her display would win first place. This 1915 display of rows of apple boxes at right shows the harvest of growers W. N. Couch, C. H. King, and Hallberg Ranch. (Above, WSCHS; at right, WSB.)

The Kauffman and Arnold apple exhibit above displays some of their Gravenstein harvest of 1910. Melvin C. Kauffman and his brother-in-law John W. Arnold joined forces to produce one of the best orchards in the county. In 1924, their orchard of 26 acres produced 5,000 boxes of shipping apples and 4,000 boxes of dried apples. This windmill from the 1912 apple show was made from the apples of D. U. Toffelmier of Sebastopol. Although the show was an incredibly popular event, the Gravenstein Apple Fair was cancelled after 1915 with the onset of World War I. However, even without the fair, the Gravenstein was still celebrated. In 1915, the Sebastopol Chamber of Commerce published an informational booklet of the Gravenstein. (Both, WSCHS.)

Sebastopol would not see the Gravenstein Apple Fair again until 1928, when it was revived by the Sebastopol Post 39 of the American Legion, a veterans service organization. With the revival came a new name—the Gravenstein Apple Show Classic. The fair was still held in the same field on Main Street, but over the ensuing years, it grew to include apple-packing contests, live music, a fashion show, dunking for apples, and apple queen contests. These two 1930s photographs show the American Legion display under the show's tent (above) and bathing beauties bobbing for Gravenstein apples. (Above, ET; below, WSCHS.)

The Gravenstein Apple Show Classic was moved to the Sonoma County Fairgrounds in 1932 by the American Legion and became part of the Sonoma County Fair until 1938. Although now part of the fair, the apple show still held on to its history by having incredible apple displays and allowing local businesses to display their products and services. This photograph shows an elephant made of apples, created by the Sebastopol Chamber of Commerce's 20-30 Club. As with other displays, the elephant sculpture demonstrated amazing attention to detail. Not only is the elephant's body made of whole apples of different color shades, but the tusks were created out of sliced apples. To give the elephant life, there is even a whole apple sitting at the bottom of its trunk as if it is ready to eat it. (WSB.)

Local businesses from various industries throughout Sonoma County also displayed their products at the Gravenstein Apple Show Classic. Lawrence Boswell, owner of L. Boswell Poultry Remedies, is standing in front of his display of products at right. A portion of Metcalf Hardware is seen to the right; they sold plumbing supplies, sheet metal, and electrical products. Regal Amber Lager Beer also had a display in the 1930s, shown below. The Regal Distributing Company had a warehouse in Sebastopol where they were distributors for Hiram Walker and Sons, Inc. Regal distributed whiskies, gins, and cordials for Walker as well as their own lager beer. (Both, WSB.)

In 1939, the Sebastopol Chamber of Commerce moved the fair back to the original site on Main Street. With the move, the chamber also renamed the classic to the Sebastopol Apple Show. By 1941, the Sebastopol Apple Show was beginning to look more like the Gravenstein Apple Fair of today. Since the first fair in 1910, it had slowly moved to the month of September. Now it included dances, music, parades, diving and swimming exhibitions, carnival acts, a rodeo, and more. However, as the United States became more involved in World War II, the show was again stopped. It would not be started again for another 30 years. This 1910 display of a rooster is titled "Cock-A-Doodle-Do" by the Cunningham growing district. In the background, apples with the Dickie Boy brand are displayed. (ET.)

In 1971, the apple show was once again revived, this time by Thomas John Symes and the nonprofit Sonoma County Farm Trails. Farm Trails was created to market the area's agricultural products through a road map of member farms. Its members decided to use the fair as a fund-raising event with proceeds going towards the creation and printing of the Sonoma County Farm Trails Map and Guide. Proceeds were also used to fund promotional and educational programs within Sonoma County. With renewed support of the event, it was given yet another new name, the Gravenstein Apple Fair. With a new name, the fair would also be moved to a new location, the Enmanji Buddhist Temple. With the success of the fair, the Apple Blossom Festival and Parade also began, but was held in the spring. Locals at the fair were admiring a large apple float on a truck flatbed on the Enmanji Temple grounds. In the background is Enmanji's Memorial Hall, built in 1953 to honor three fallen Sebastopol soldiers from World War II: Leo Kikuchi, Peter Masuoka, and Joe Yasuda. (DRC.)

Creating apple sculptures even became a family affair, as seen in these photographs of locals Gloria Roberts and her children Scott and Carrie in 1974. On behalf of the Sebastopol Apple Growers Union (SAGU), the family was creating a display of Johnny Apple Seed. Johnny's body was made of pepperwood and oak, with the face made using a chain saw. The finished sculpture held a Bible in one hand and a bag of apple seeds in the other. On his head, a tin pot was used for a hat. Unlike fairs of the past, this sculpture was created on the back of a flatbed truck for a parade. Don and Gloria Roberts' family has grown Gravensteins for four generations on the same Gold Ridge area property. (Both, DRC.)

There would be two more location changes in the ensuing years. In 1976, the Gravenstein Apple Fair was held at the Forestville Youth Park and called the Sebastopol Apple Fair. In 1977, it was moved to its current location at Ragle Ranch Park in Sebastopol. Still, the fair has continued to celebrate the local agricultural life as the original fair committee set out to do. Most importantly, no matter how many times the fair had to be reorganized, it maintained its history of apple sculptures as seen in these two photographs of an apple robot (at right) and a large apple (below) created in the 1980s. (Both, WSCHS.)

With a long history of economic importance, Sebastopol can never forget the fruit that put it on the map. Even without the elaborate displays and exhibits of the early 1910s, today the Gravenstein Apple Fair still draws over 16,000 visitors each year and remains a family affair. As with the first fair in 1910, the fun fair atmosphere continues with live music, wagon rides, cooking shows, beekeeping, and animal demonstrations. Now more than 130 artisans display their crafts, which include art, jewelry, handicrafts, local wine, cider, and of course apple pie. Still, those grand exhibits of 100 years ago are unforgettable and remind people of a time gone by, as witnessed by these two 1912 prizewinners—the Flying Dutchman (at left) and Ferris wheel. (Both, MA.)

Two

THEY STARTED IT ALL

When the first immigrants settled in Western Sonoma County's Green Valley, they recognized that the soil and climate were ideal for growing fruit trees. Early settlers from the East and Midwest were often experienced in farming. They applied those skills, or quickly learned new ones, to prepare and till the soil and plant their first crops. Most planted family orchards with pears, cherries, and apples, as well as gardens. The orchards and gardens thrived, as did the families. Agriculture in the area continued to thrive and new farming techniques and equipment were developed. Improvement in roads, trucks, and trains increased the quality of transportation of fruit. Many of those early ranchers held on to their land, improved it, adopted new ideas and techniques, and passed their land on to their children. Hard work and perseverance were required to meet the challenges encountered by each generation. Little did those early orchardists realize that as many as five generations later, their land would continue to produce, among other crops, outstanding apples.

This unidentified rural Sebastopol scene is typical of the farms established in the mid- to late 1800s. Oak trees have been cleared and an orchard planted. A well-tended vegetable garden and pens for chickens, pigs, and cows round out the accommodations for this sustainable family farm. They had ample water, homegrown fertilizer, and firewood for heating and cooking, which was all that was needed for a dedicated, strong, and determined family to succeed; countless families did succeed in those early years. (WSCHS.)

In 1850, Isaac Sullivan and his father-in-law, Mitchell Gilliam, planted 150 apple trees that they purchased for $1.50 each from a Petaluma nursery. This orchard was one of the first planted in the area. The Gregsons, Marshalls, Gilliams, Winklers, Purringtons, Thomases, and Churchmans were among those who planted trees in the Green Valley area (later to become Graton), north of what is now Sebastopol. (HN.)

Clayton Winkler built his home in Green Valley in 1860. The family home stood until 1920 when it was razed and much of the lumber used for the construction of a Spanish-style bungalow. Clayton is said to have planted his orchard from stock sent from back east. His son Walter Scott Winkler, born in 1883, later began selling bare root trees. He started the Winkler Nursery on the property, because he felt local nurseries were too expensive. He never advertised. A neighbor, tiring of answering queries regarding the nursery, eventually put up a small wooden sign. Walter moved onto the home place prior to his father's death in 1968 and ran the nursery until his son James took it over in the 1980s. He is quoted in the *Winkler Family History* as saying, "The family nursery was a hobby of my father's, but it has developed into a good portion of our gross income." The three-generation business operated for 46 years. Clayton Winkler died in 1901, and his wife, Martha, in 1905. They are buried together at Sebastopol Memorial Lawn Cemetery. (WSCHS.)

John and Neta Hallberg, Swedish-born parents of Alfred and Oscar, settled in the Sebastopol area on the 40 acres that John had purchased in 1880. He built one of the first apple dryers in the area. It was destroyed by fire in 1905. At one time there were 101 dryers or dehydrators between Forestville, Cloverdale, and south to Hessel. All but 15 burned. (DRC.)

Alfred and Oscar Hallberg started drying apples in 1916. They built one of the first floor-type kiln driers in Sonoma County. Oscar purchased apple orchards until he owned 500 acres and leased more. He devised a method of making and canning applesauce and purchased the Green Valley Cannery. His cold storage plant was the first in the county. It provided a convenient market for local apple growers and had a capacity of 100,000 boxes. (WSB.)

Walter Scott Roberts left Iowa and settled in the Sebastopol area in the late 1880s. He cleared the forested land of oak, fir, and redwood trees. A donkey engine was used to drag away the felled trees. Itinerant men would work with breaking plows, pulled by big teams of 10 or 12 horses, to grub out the stumps and roots. Roots left from oak trees could cause oak root fungus that would kill apple trees. Only when the roots were eliminated would the farmer plow and plant the field. Sammy Thomas is seen plowing with Scott Roberts's team on Roberts's Gold Ridge land in preparation for planting an orchard. The young Walter Roberts family is seen below in the early 1900s. Young son, Ivan, (middle) carried on the apple farming tradition on the Roberts' land. (Both, DRC.)

Ivan Roberts cleared his own land and planted an orchard when he was 18 years old. Harvesting apples was a family affair, and child labor laws were lax. It was a common sight to see a small child napping under a tree while his parents and young siblings picked apples. The wife often worked in the orchard, dryer, or packinghouse after her household chores were complete. Ivan (at left) is shown harvesting his crop around 1935. A few years later, his wife, Gertrude, and young son, Don, are seen during harvest (below). The apple-picking bag on her shoulder is not a fashion statement. The orchards of Walter Roberts were passed on to his sons, Ivan and Bill, and in turn, their sons, Will, Don, and Lloyd. Lloyd's son Randy is the fourth generation to work the land cleared and first planted by Walter Roberts. (Both, DRC.)

George Huntley moved to Sebastopol in 1880. He bought 80 acres and planted 40 of it in mixed fruit. While his apple trees were maturing, his first crop of berries netted $11,000. When Huntley died in 1901, his 125 acres bordered Burbank's farm on the south, and Valentine Street on the north, from Pleasant Hill Road to Jewell Avenue. The Huntley Packinghouse, shown in 1900, was destroyed in the 1906 earthquake. (RC.)

Andrew Frei came to the United States from Switzerland in 1852. He manufactured furniture in San Francisco until, in 1882, he had enough money to buy land in Sonoma County. Later his sons Louis and Walter, having inherited his Frei Road land, owned one of the largest orchards in the county. Men and women are seen working in the Andrew Frei dryer around 1910. The worker on the right is Erroll Singmaster. (HSJ.)

Nathaniel Griffith (at left) planted his Laguna Road orchard in 1883. He worked with Luther Burbank (below) in 1890 to identify the unknown cultivar growing on Griffith's land. The two developed the Gravenstein apple. Griffith's was the first successful commercial Gravenstein apple orchard in Sonoma County's Gold Ridge area. He was shipping apples in carload lots before other orchardists realized their value. He was aptly named "Grandfather of the Gravenstein." A lover of nature, Griffith preserved 10 acres of his property. That land, Griffith Woods, is still available to the public today for special events. (At left, CF; below, WSCHS.)

In 1892, Thomas E. Barlow originally purchased 35 acres of land in Green Valley. He raised blackberries as well as apples. Thomas gave the right-of-way for the Petaluma and Santa Rosa Railroad to cross his ranch. Barlow station was named in his honor. He died at age 34, leaving his widow with six children and 600 acres of apples and berries. Laura Barlow continued her husband's ranching policy and aptly carried on his work until her sons joined her. Those sons, Thomas Denman Barlow Sr. and Leland Barlow, pioneered applesauce canning in 1939. Women are shown at work in the first commercial applesauce cannery. Laura Barlow is pictured in a family photograph around 1910. The Barlows shown below are, from left to right, Louise, Leland, mother Laura Barlow, Bill, Maude, Denman, and Bess. (Both, TDB.)

Arthur Upp, a San Francisco native, was born in 1876. He graduated from California College of Pharmacy in 1897. He owned a successful drugstore in Graton. In 1910, he planted the family orchard on the road west of Graton that now bears his name. In 2010, there are still 70 or 80 original Upp trees on the property known now as the "Walker Ranch." Lee Walker pointed out one of those original trees still bearing fruit, as the Walkers celebrated the 100th anniversary of that tree. Arthur Upp was the maternal grandfather of Lee Walker. The next two generations of Lee Walkers still operate the ranch. Lee and wife, Shirley, have been innovative in marketing their products. In their early years, they sold apple cider from a roadside stand. Lee Walker II is shown sorting apples above. (LSW.)

John Walker, Lee's paternal great-grandfather, was born in 1826 in Missouri. John was one of the original settlers in the Sebastopol area, arriving in the 1840s. He built the first redwood house in the area and partnered with Joseph Morgan Millar to open the earliest mercantile establishment, a trading post, in 1849, south of present-day Sebastopol. He had a general interest in farming and stock raising. At the time of his death in 1895, he owned 4,000 acres. His son, Edward, born in 1866, upon reaching his majority was given 529 acres of the home ranch. His two sons, Edward Lee and John, eventually inherited his property. In turn, Lee's sons, Lee II and Arthur (Bob), inherited the Walker orchards. The two agreed that Lee would operate the ranch of the Upp grandparents and Bob the Walkers' property. (MF.)

In 1920, Rudolph and Maude Oehlmann established a small dryer known as the "Oehlmann Evaporator." They operated the business at the same site in Graton for 25 years and then incorporated and changed the name to Manzana Products. The company took its name from the Spanish word for apple (*manzana*). Oehlmann was active in the business of "the apple" until his death in 1955. Maude and Rudolph Oehlmann are pictured. (Both, MZ.)

In 1948, Manzana Products had a catastrophic fire. Employees attempted to put out the fire with their cider vinegar. In spite of their efforts, everything was lost. The Oehlmanns rebuilt after the fire, combining dryer and cannery for the first time. Soon after that time, the dryer was discontinued. The Oehlmann's sons-in-law currently operate the company; Howard Norton married the former Edith Oehlmann, and Ralph Sandborn married the former Constance Oehlmann. Both sons-in-law joined the business after completing military service. During the height of each season, Manzana Products employed approximately 100 women and 25 men. The firm also leased and operated 100 acres of apples. (MZ.)

William Hotle, in partnership with Lou Hart and George McNear, established the 200-acre Gold Ridge Ranch and packinghouse. McNear served as president of the company. The Hotle land was considered by many to be the best fruit ranch in the country. After Hotle's death, his orchards were passed to his daughter Miriam Hotle Burdo; she continued the family tradition as a grower. During the war, many Japanese apple ranchers were relocated leaving their property in Sebastopol. It has been estimated that 70 to 80 percent of those interned returned to the Sebastopol area. Many had Caucasian friends and neighbors who stepped in, ran their ranches, and harvested and sold the crops so that taxes could be paid. William Hotle cared for the Furusho property. Don Lunceford, Jim and Don Williams, Fred Miller, and Joe Cordoza were also among those helping their neighbors. (DRC.)

Joe Furusho was born in 1916 on the Hotle Ranch where his father worked. As a youngster he labored with his father. When his father purchased his own ranch on Burnside Road, Joe helped. During the Depression, in order to help keep the family property, Joe quit high school. After Pearl Harbor, he continued farming until he was evacuated to Colorado. Joe and his wife were the first couple married at Amanche internment camp. While in camp, he leased (financed with references from people in Sonoma County) and farmed land in Colorado for two years. He saved his earnings, and with those proceeds he was able to hold on to the family property in Sebastopol. A farmer and businessman, after the war, Joe and his brother Tom owned and operated Furusho Brothers Farm and Packing Company in Sebastopol for 50 years. The Jackie Boy label was one of theirs. Charles Yamamoto contracted with the Furushos to operate their dryer. That became a family operation, with his wife and five daughters working full time. They began during Easter vacation, and apple work continued all summer. (WSCHS.)

In 1917, Kichizo Morita, at age 14, came by himself from Japan to the immigration checkpoint at Angel Island. He attended photography school and opened a studio in Salinas. In 1937, he bought (in his wife's name) 23 acres of land on Sanders Road from the Elphick family. He and his wife cleared the land without machinery and planted apples with berries between the rows. Eventually he built a packinghouse. Florenzio and Julia Lorenzo cared for his property during the war years. As was the case with most apple ranchers, his wife and children all worked in the harvest. A woman who had spent her adult life helping with the family harvest in Green Valley, referring to the rigors of that life, was quoted as saying, "When I die, if I go to a place where there are apples I'll know it won't be heaven." The Moritas would all agree that apples mean hard work. (KF.)

Joe and Elizabeth (Lea) Morelli Perry had moved in 1930, as young people, to the Perry family's Twin Pines Ranch on Gravenstein Highway North. Lea (shown above) worked outside the home at Analy Saving Bank (later Bank of Sonoma County) and on the ranch. The Perrys had 44 acres of apples and farmland with animals and an apple dryer. The back 32 acres was later sold to the Sebastopol Co-op Cannery so that their excess industrial water could be sprayed onto the newly terraced area and then pumped into a holding pond before emptying into the Atascadero Creek. The operation of their ranch was a full-time job, but during the war years they made time to care for the Covey Road property of the Kaz Ito family, site of present-day El Molino High School, and the Johnny Ito farm on Hurlbut Avenue. Lea Perry was given power of attorney and handled most of the Ito family paperwork while writing detailed letters to the owners in the internment camp, reporting on their property and business interests. She did all this while working exhausting hours driving a truck between the farms, driving a tractor, and supervising work crews during the harvest season (TP.)

Jim and Walter Taniguchi, like many apple growers, operated apple dryers on their property. In 1972, the brothers shared the role of grand marshal in the Apple Blossom Parade. That same year, Amaya Taniguchi (at left), Jim's wife, won top honors for her apple pie recipe. Popular as a cook, she was also accomplished in crochet work, flower arranging, and "Utai," a Japanese style of singing. All of that was after Amaya Taniguchi completed her day in the orchard. The Taniguchi dryer (above), stood near Bloomfield Road. (Both, RTF.)

A Gravenstein tree on the property of John W. Arnold became well known locally for its heavy production. The record was set in 1935 when 123 boxes, at an average of 41 pounds of fruit per box, were harvested from this tree. The tree was planted about 1893 by Frank Kauffman and had been bearing for 34 years when this photograph was taken. The Kauffman exhibit at the apple fair displays their fruit. Members of the Arnold family are shown, in 1929, in front of their famous tree. Pictured are, from left to right, Minnie I. Kauffman Arnold; her husband, John W. Arnold; Meta Arnold; and her husband, E. W. Arnold; and Vivian E. Arnold. (Both, WSCHS.)

Darrel Hurst and wife, Margie, were married in 1942. That same year they purchased 40 acres of apple orchards. The next year, Darrel enlisted in the U.S. Marine Corp. It was two and a half years before he returned to his wife and young son and apple ranching. He acquired more land and more children. The couple raised 13 children, including eight who were adopted. An apple farmer for 55 years, Darrel was known for being innovative in fashioning tools and equipment. He opened a gift shop at his Twin Hills Ranch, where local products and apple pies and bread are sold. His family continues in the business. Darrel is shown with a load of Hurst apples ready for market (above) and at the first "Apple Blossom Time" celebration at the Emanji Temple with his prizewinning apple sculpture. (Both, DRC.)

George and Jean Focha Mennini worked their 500 acres of apples in the Gold Ridge. After Jean's death, George carried on alone. He worked sun up to sundown, but often, in the heat of the day, he would park his tractor under an apple tree and nap for a couple of hours before going back to work. He died when he was well into his 90s on his John Deere tractor. (DRC.)

Harry and Candida Silva settled in the Cherry Ridge-Grandview area in 1922. Their son Bill is shown above preparing the mixture of sulfur and lead arsenic used for spraying his orchard. Silva Brothers packinghouse was started in 1954 when Bill and his brother Jess bought out their father. They hauled their own apples—fresh, dry, and candied (in their plant). In 1958, they became commercial carriers as Silva Brothers Trucking. Bill continues, in 2010, to drive long hauls. (BS.)

Apple juice stands popped up along county roads; farmers found that to be another way to sell the family product. The Rosebrooks contributed to that tradition. William and Leona Rosebrook moved to the Sebastopol area in the early 1900s. Rosebrook developed the Rosebrook Gravenstein, which is still a mainstay with local orchardists. William's brother Harry is seen to the right of the family fruit stand on Gravenstein Highway North (above). They were among the first to offer frozen cider, a real summer treat. Not far from there, on the west side of the highway, in 1953, Marvin Zimpher opened his fruit stand. It occupied the site until 1980 when Mary's Pizza was built on the old Zimpher property. Young Eileen Keyser Zimpher is seen (below) with Zimpher apples in 1957. (Above, RF; below, ZF.)

Apple cider, a favorite side product of the apple harvest, requires a great deal of time and preparation. From the earliest apple harvests to the present day, young family members have contributed to the apple workforce, and making cider has become a family activity. It can be messy and requires a great deal of hard work but provides much satisfaction. Geoffrey Skinner (at right) began at an early age, in 1968, washing apples for the family's cider production. Seventeen years later, he was still washing apples before adding them to the cider press. (Both, LSC.)

Many farmers who raised apples settled the western Sonoma County hills. Those pioneers were joined through the years by many more orchardists, those with hundreds of acres of apple trees and those with small orchards. A limited number of family names are being carried on by the current generation, in some cases, five generations later. Eric Marshall, third in his family line, continues to harvest apples on his family property and maintains orchards for other ranchers. He is shown in the orchard with his father, Tom Marshall. Apple growers were many in number: Alves, Baldi, Batten, Best, Bollinger, Braga, Briggs, Cardoza, Deyer, Doty, Douglas, Duckhorn, Elphick, Garozzi, Germone, Gregori, Gudino, Gustafson, Hankins, Hastings, Heintz, Hillard, Hunt, Jewell, Kelly, Kikuchi, Kozlawski, LeBallister, Lunceford, Martin, Martini, Meyers, Morgan, Nelson, O'Connell, Orr, Pelletti, Perroni, Peterson, Ratzlaff, Ross, Rued, Sanchetti, Showwalter, Silveria, Starkey, Stone, Tabor, Tischer, Trefcer, Trigeiro, Voit, and Winsley. The list goes on and on. There are many names left unrecorded and many more untold stories of the harvest of the apple. (PC.)

Three

To Market We Go

The Sebastopol Apple Growers Union (SAGU) was founded in 1911 by William Hotle as a nonprofit organization, and it was reorganized in 1919. It became one of the major players in the apple industry in Sebastopol. By 1925, SAGU had two large processing facilities. Other major producers of apple products included Speas Company, Garcia and Maggini, Oehlmann Cannery (later renamed Manzana), the Barlows, the Sebastopol Cooperative Cannery, and Vacu Dry. After 1900, many apple growers turned to cooperative canneries to process their product. Due to the high cost of packing and shipping, throughout the second half of the 20th century, the cooperatives played a primary role in the industry.

While most workers hailed from local areas in West Sonoma County, some migrated to Sonoma County from Japan, Mexico, the Philippines, and Jamaica. Workers harvested from mid-summer until the end of September, annually. Such workers often stayed at migrant camps such as the one at Handy's Corner or at the campground adjacent to the Laguna watershed within walking distance to the Sebastopol Cannery or Speas Company. Many, not all, seasonal workers lived in surrounding communities. They were efficient and punctual, hard-working folks who valued their jobs despite most earning low wages. Before 1941, most earned less than 40¢ an hour and worked in excess of 10-hour days. Before World War II, the apple workers did not possess the advantage of union representation. In late July 1935, two Santa Rosa activists, aided by a group of the youngest workers in the processing plants, briefly struck the Sebastopol Apple Growers Union. After the mob of so-called "vigilantes" brutally abused the most vocal union leaders, the strike failed. Ultimately in 1956, following a brutal beating of union organizer, Bill Grami, and a subsequent national boycott of apple products, the orchard owners and officials in the cooperatives recognized the Teamsters Union. With the union came the eight-hour day in the dryers and canneries as well as equal pay for equal work for laborers regardless of sex.

July is the Gravenstein apple harvest time in Sebastopol. William Borba documents the scene here in the early 1900s. Borba was a local businessman very interested in history and photography. Originally from the Azores, the Borbas came to Sebastopol in the late 1800s. The family patriarch, Manuel Borba, worked in local orchards pruning. The family used the prunings in their home stove for cooking and heating. William Borba's documentary photography included many scenes from the early apple shows and yearbook photographs from Analy High School. Some of his glass negatives now reside with the Western Sonoma County Historical Society. Borba was one of the founders of the Sonoma County Historical Society in the early 1960s. (WSB.)

Al Focha applies fungicide to dormant apple trees at the Perry's Twin Pines Ranch in Sebastopol in 1910. Sulfur and lead arsenic were common chemicals used to control fungus and insects in the first half of the 20th century. Manuel Perry pauses with his horse-drawn harrow used to smooth the plowed ground in his orchard. All farm machinery in the early part of the 20th century was horse drawn, since gasoline-powered farm equipment was not readily available or affordable to Sebastopol farmers in those days. Farmers often helped neighbors by lending their equipment and their labor. (Both, TP.)

Although there were numerous apple-processing sites in Sebastopol, Graton ably competed in the apple industry. The arrival of the Petaluma and Santa Rosa Railroad in 1904 paved the way for easy shipping of fruit. Many hands were needed to pack fresh apples and quickly get them to market before they went beyond the peak of ripeness. Modern, mechanically refrigerated railroad cars were not employed until the late 1940s. The Pacific Fruit Express railroad cars carrying freight out of Sebastopol originally used ice to chill the cars. They operated 18 artificial ice plants, the largest located in Roseville, California. They harvested over one million tons of ice annually. Fresh fruit being shipped needed to be chilled, using a 2-inch layer of ice placed on top of the load. It was found this method was ineffective, as the ice only cooled the top layers of fruit and it had to be re-iced daily en route to its destination. A Pacific Fruit Express boxcar resides at the rear of the West County Museum in Sebastopol and serves as the historical society's archive storage. (DWD.)

Around 1900, workers gathered at plant no. 3 of the Central California Cannery in Sebastopol. Seasonal cannery work supplemented local incomes. The third person from the left is Hubert B. Scudder, who became Sebastopol's mayor in 1929. The Central California Company had operations in several northern California fruit-producing areas. In 1916, they merged into the California Packing Corporation, or CALPAK, with headquarters in San Francisco. They controlled a vast network of drying and canning operations and had the financial resources to create a national market for their products. Their premium label was Del Monte. Drying fruit could be done in the sun in central California, but Sebastopol's varying weather necessitated heated dryers. Canning operations required more investment. Early canneries used the assembly line style of work to peel, slice, pack, and cook fruit. Then the canned goods were solder sealed, the labels affixed, and finally the cans stored until shipped. (WSCHS.)

Many apple farmers had their own dryers. This Occidental Road dryer, photographed in 1960, is of typical wood construction. Total destruction by fire was a common occurrence. The Perrys and extended family pose in front of the first apple dryer on their Twin Pines Ranch around 1910. From left to right on the porch are Manuel Rose, Ann Perry, Emma Focha, unidentified, and Mary Perry. Matriarch Anna Perry stands on a platform below the porch, and her husband, Manuel, sits with the driver on the wagon. Apple-drying methods changed over the years. The first dryers were fueled by wood and eventually natural gas. The process evolved from turning the apple slices by hand to their being placed on a moving belt that ran above a natural gas–powered heat source. The process went from taking days to hours to dry apples. (Above, DRC; below, TP.)

Operations at the Pleasant Hill Dryer included an apple-grading machine. It would depend on the amount of pest damage whether the apple was graded no. 1, no. 2, or juice. Pleasant Hill's staff members Ted Maddocks (left) and Clarence Pete take a lunch break below. Pete was manager of the Pleasant Hill Dryer. There were many cooperatives formed to market dried apples alone. Growers banded together to set prices that were equitable and permitted them to make a profit. The Pleasant Hill Cooperative Dryer, the Molino Apple Cooperative, the Green Valley Dryer, the Cherry Ridge Apple Processing Cooperative, the Sebastopol Processing Cooperative, and the Forestville Processing Cooperative were supported by a central sales and accounting organization called the Analy Marketing Cooperative. (Both, DB.)

The Sebastopol Apple Growers Union had a packinghouse on High Street until the 1940s. After World War II and the decrease in demand for apple products, this facility was torn down and the operations consolidated in a plant on Occidental Road. The new facility provided a cannery, cold storage plant, packing shed, and warehouse space. Both facilities had access to the Petaluma and Santa Rosa Railroad line. The Sebastopol Apple Growers Union canned 500,000 cases of sliced and sauce apples in 1954. That was a 100,000-case increase from the previous year. Three hundred growers were members. The Sebastopol Veterans Building subsequently arose on the High Street site. (Above, WSCHS; below, MA.)

The Garbro (Garcia Brothers) orchard was located in the Vine Hill Road and River Road area. A 1910 vintage truck bears solid rubber tires, making for a rough ride on Sonoma County's early roads. Transportation of farm products had always been a stumbling block to farmers wanting to sell commercially. Until the Petaluma and Santa Rosa Railroad (it arrived in 1904) and the Northwestern Pacific Railroad (it came to Santa Rosa in 1872) approached Sebastopol, farmers had to haul goods to Petaluma to load onto ferryboats and thence to San Francisco. From the 1850s to 1900, as many as 50 ferry boats carried goods to Bay Area ports. The Petaluma and Santa Rosa Railroad owned its own ferryboat, the *Steamer Gold*. The last ferryboat to land in Petaluma was in 1950. The Garbro workers have a load of windfall apples used for juice or vinegar. Gathering windfalls is intensive manual labor with each worker gathering apples from the ground into burlap bags then sewing them shut with twine. (DB.)

The large packinghouse of the Garcia and Maggini Company in Sebastopol was located just east of the Petaluma and Santa Rosa Railroad's depot. They were in an extremely convenient location to ship their products on the railway's electric line. Garcia and Maggini, headquartered in San Francisco, operated plants in Fresno, Watsonville, and Sebastopol. From business directories, one learns that Garcia and Maggini operated in Sebastopol into the 1920s where they shipped both raw apples and evaporated or dried apples. Their workers pause for a photograph with the Bon Ton labeled boxes, a Garcia and Maggini brand. (Above, JM; below, MA.)

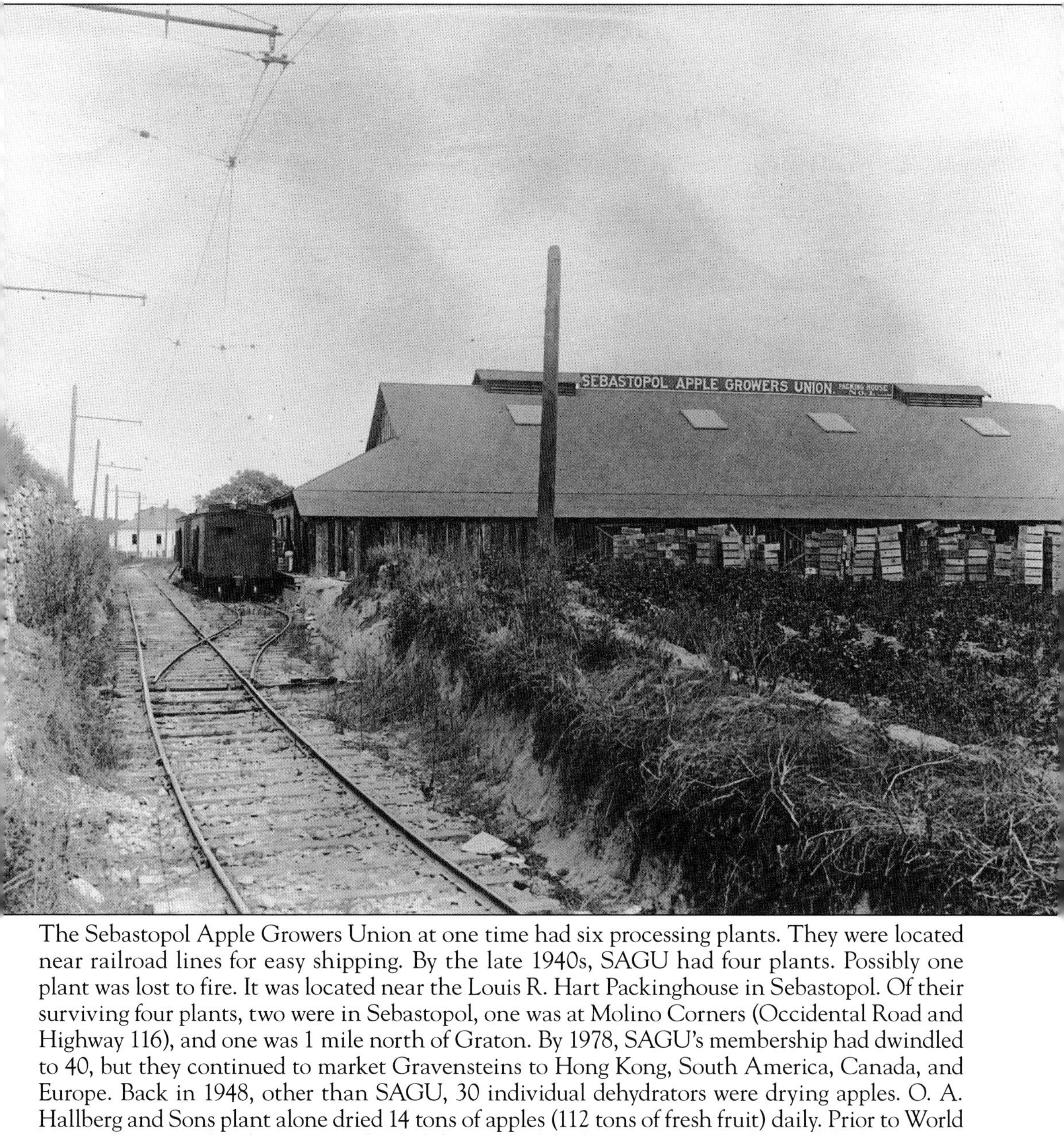

The Sebastopol Apple Growers Union at one time had six processing plants. They were located near railroad lines for easy shipping. By the late 1940s, SAGU had four plants. Possibly one plant was lost to fire. It was located near the Louis R. Hart Packinghouse in Sebastopol. Of their surviving four plants, two were in Sebastopol, one was at Molino Corners (Occidental Road and Highway 116), and one was 1 mile north of Graton. By 1978, SAGU's membership had dwindled to 40, but they continued to market Gravensteins to Hong Kong, South America, Canada, and Europe. Back in 1948, other than SAGU, 30 individual dehydrators were drying apples. O. A. Hallberg and Sons plant alone dried 14 tons of apples (112 tons of fresh fruit) daily. Prior to World War II, Japanese families operated 72 of the area's dehydrators. In 1948, the 30 dehydrators had as much, or more, production capacity as the previous 80 plants. (JM.)

In the 1930s, the Speas Manufacturing and Distillery in Sebastopol stood on the corner of Petaluma and McKinley Avenues. Built originally in 1919 by Frank Butler, the plant was later sold to Speas and Butler continued on as manager. After Butler's death, his secretary Virginia Sustarich managed the plant, which employed 50, for the next 20 years. When demand for vinegar diminished, they expanded to distilled apple brandy and apple wine. It took about 2,000 gallons of juice to produce 40 gallons of brandy. A by-product of apples was PEN-JEL, apple pectin used by homemakers to make jam and jelly. The large, well-built concrete structure on the right was the bonded warehouse. In 1936, Speas also made whiskey with corn from Argentina, rye from Oregon, and malt from Minnesota. The company disbanded in the 1980s, and the warehouse was eventually converted to the Sebastopol Cinemas in 1993. (Above, WSCHS; below, DM-SCL.)

Rudolph and Maude Oehlmann operated this 10-kiln dryer in Graton, one of the largest in the area. Fire destroyed it in 1948. They renamed the business Manzana and concentrated more on canning than drying. In making applesauce, workers chose larger-sized apples. Small apples were too difficult to peel mechanically and so were used in juice. Once apples are peeled, they are heated, and the pulp is pressed through screens. Before it is canned or bottled, workers remove any remaining unwanted apple parts. Manzana uses the brand name North Coast, and the organic labels are Solana Gold and Crystal Wave. In 1953, Manzana packed 350,000 cases of applesauce. They had just installed equipment to can sliced apples that year. (Both, MZ.)

The Furusho brothers, Tom and Joe, operated a packinghouse located off South Gravenstein Highway behind the current McDonalds building. In the packinghouse in the 1920s, men assemble boxes of the Furusho's Jackie Boy and Everyday brands. Different boxes were used in different ways to handle apples. Many farmers made their own lug boxes from basic parts called shook. Glaser's was the main supplier of shook in Sebastopol. Field lug boxes were used for harvesting. They hauled apples to packinghouses, canneries, and dryers in lugs. They were usually stenciled with the owners name and reused for many years. Sometimes box makers traveled the harvest circuit. They carried their own tools and assembled shipping or one-way lugs for packinghouses. A good box maker could assemble up to 500 one-way lugs in a day. After the 1950s, cardboard boxes replaced wooden ones. (Above, WSCHS; below, SCL.)

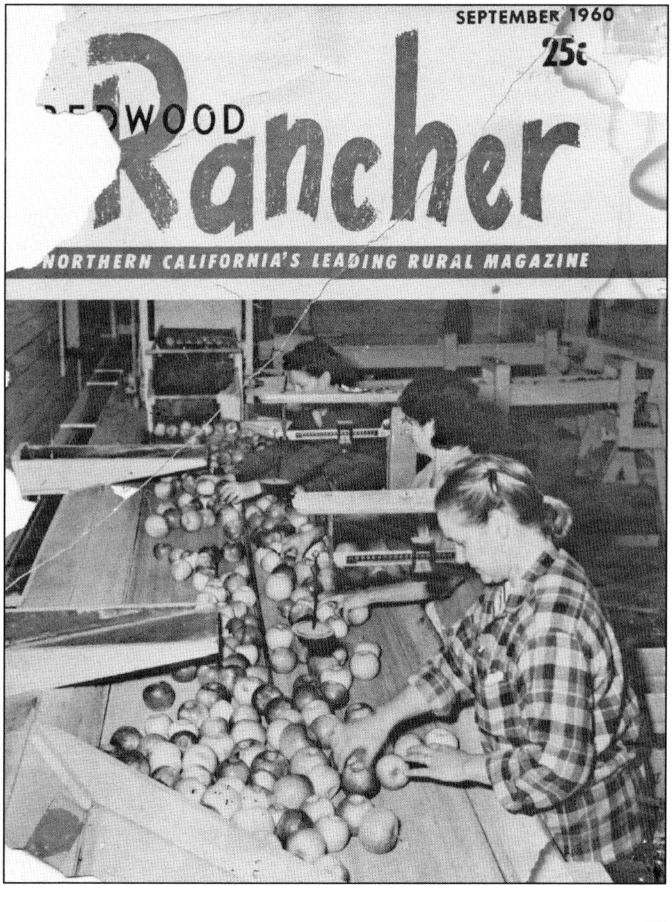

The Silva family orchards and processing operations were located on Cherry Ridge Road. Above, in 1937, with the dryer in the background, are, from left to right, Candida Silva, son Jess (in front), along with son Bill Silva, and Harry Silva. The Silvas' operation was featured on the cover of the *Redwood Rancher*. The three workers cutting and packing apples at right are, from front to back, Mrs. Woods, Candida Silva, and Mrs. Packard at the Silva Brothers Ranch. When the apple market slackened, Bill Silva used his trucks to haul apples for other growers and it grew into an active trucking business. The Silvas found a unique outlet for their apples one time when they shipped apples to Thailand to be used to feed monkeys. (Both, BS.)

The Barlow family were pioneers in canning applesauce at their home operation in 1939. They expanded into a plant in downtown Sebastopol that continued operations until recent years. Charlotte Barlow O'Brien, below, checks in with the work crew at the Forestville Cooperative plant where she was manager. There is some serious apple-pressing going on by the size of the equipment. Apple concentrate was used to enhance the flavor and aromas lost in jam and jelly making. The Silveira and O'Connell processing plant even produced an apple essence product, which was a water-like liquid. The apple pulp that remained after the extraction process was hot air dried and became pomace, which could be fed to cows. (Both, TDB.)

An aerial photograph dated from 1955 shows the extent of the Oscar A. Hallberg Cannery and Cold Storage buildings in Graton. The photograph was taken north to south and includes a portion of the town of Graton together with surrounding countryside. Oscar Hallberg began by working on his father, John's, apple ranch, then purchasing neighboring acreage before he married his first wife, Mary E. Barlow. Oscar Hallberg (1893–1976) is credited as being the first to replace wood with oil and natural gas to fuel the apple dryers. The annual output of dried apples averaged 300 tons. Hallberg's large cold storage building held 100,000 boxes, the largest such building in Sonoma County. Hallberg was a member of the California Apple Council. (TDB.)

Vacu-Dry employees work a slicing machine in 1960. Vacu-Dry was incorporated in 1946. Its sole business was in drying apples. Dried apples can be infused with other fruit flavors and used in cereals and cake mixes. In 1999, Vacu-Dry closed and laid off 125 employees—some had worked there over 30 years. The Vacu-Dry directors planned to sell some equipment along with the brand but not inventory or real estate. They reinvented themselves as Sonoma West Holdings, Inc., and planned to lease out their premises to other business operations. Vacu-Dry's closure in 1999 left many growers in chaos with harvest season approaching and no place to take their crop. An increase in apple orchards being converted to vineyards was predicted. (Both, DB.)

Four

Here Comes the Parade

The Gravenstein apple has been celebrated in many ways in Sebastopol. There were the early apple shows with large, elaborate sculptures constructed of fresh and dried apples. A variety of apple products were featured. In later years, there were additions such as live music, fashion shows, and apple bobbing. For a period of time, the site was moved to the Sonoma County Fairgrounds and then back to Sebastopol. The most long-standing celebration of the apple has been the events honoring the apple blossom. The late George O'Keefe, a local realtor, organized the first apple blossom festival. Called Apple Blossom Time, it was held the last weekend in March and the first in April 1947. The event consisted of a 20-mile tour and a flower show. The chamber of commerce sponsored bus rides into the countryside where apple growers posted homemade signs to mark the way. Alfred Hallberg was the one who erected signs to guide the postwar motorists along the route through the Sebastopol Hills. Apple blossom trail maps were seen yearly until the early 2000s. Apple Blossom Time observance evolved into a two-day festival scheduled to hopefully coincide with the blossoming of the apple trees. The chamber of commerce endorsed the event as a way to entice visitors to the area and to promote local apple products. Whatever the venue or theme, the people of Sebastopol and outlying communities have always offered their hearty support. By 1950, service groups and other sponsors were added, and the festival was renamed "Apple Blossom Festival." Additional events have come and gone, but the parade persists.

George O'Keefe's original ideas expanded and changed with the times. Ed Trigeiro came up with the idea, and a parade was added in 1968. The Sebastopol Chamber of Commerce has been a consistent sponsor. Chamber-sponsored bus rides through the blossom-laden countryside disappeared long ago, and in recent years, due to loss of orchards to vineyards, there have been no apple blossom self-guided tour routes. The self-guided map directed visitors to the orchards and back to Sebastopol. George O'Keefe was the founder of the first Apple Blossom Time festival. (At left, WSCHS; below, SCC.)

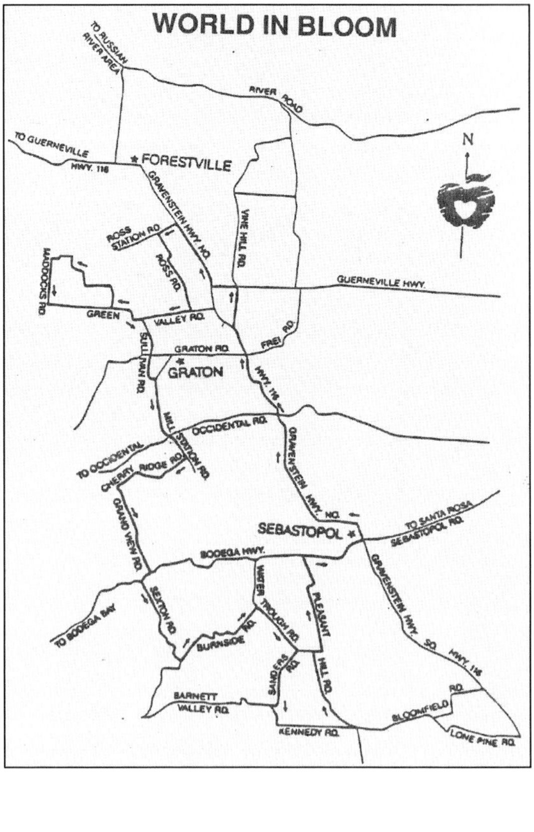

Parade traditions changed with time; there have been additions and deletions, but some entries have remained constant. The honor guard of Sebastopol's Gold Ridge Post 3919 of the Veterans of Foreign Wars always leads the parade. The late Jim Nagy, local historian and well-connected citizen, is shown carrying the American flag. Soon after the honor guard comes the automobile carrying the parade's unidentified grand marshal, below. That honor has been bestowed on a variety of citizens. Charles M. Schulz and Karen Valentine were among those who have evoked widespread recognition. (Above, RS; below, SCC.)

Tim Moore was recognized as citizen of the year in 1997. As an honored, prominent citizen, he was assured a ride in an outstanding automobile. That weekend in April was sunny and dry, which was not always the case. Through the years there have been dreary and rainy parade days. Rain tends to dampen the participants but not the spirit. Grand marshal was not the first Apple Blossom Parade role of note played by Karen Valentine. She was unable to appear in her first role of a princess attendant. In 1953, Karen and Bonnie Staple were the chosen attendants to princess Sue Edwards by a vote of older students. Unfortunately Karen missed the coronation when she came down with the measles. (Above, RS; below, DRC.)

Five-year-old Jeannie Crider was the first apple blossom royalty, voted queen in 1949. Queens and princesses, through the years, were selected by different standards and by a variety of judges. There was no selection for the five years following the measles episode. During those years, the festival was referred to as "Easter Time is Apple Blossom Time in Sebastopol." After the five-year-olds, high school seniors became the candidates, with the queen being determined by ticket sales. In 1961, a talent show was instigated, judged by a panel of community members. The 1969 queen, Kathy Hug, is shown above with her attendants, and Kathleen Skinner, shown at right, served as queen in 1966. Jennie Rivera was the last crowned queen in 1989. Enthusiasm for the pageant waned, and a transition was made to "Youth Ambassadors," a scholarship and education program that lasted for three years. (Above, SCL; at right, LSC.)

After all the decisions and work, when the day arrived it was fun to be in a parade. One got a new perspective of the organization, the event, and the city of Sebastopol from atop a float. The Saucy Squares Express carried a wooden platform for a quick do-si-do. Luther Burbank was played by Ray Kraft in the Western Sonoma County Historical Society entry. Burbank is honored and remembered for the years of plant research he conducted at his Gold Ridge Experiment Farm on Bodega Avenue in Sebastopol. The farm is maintained and operated by volunteers of the historical society. In both photographs, note the crowds, situated for long, comfortable mornings. (Above, DM-SCL; below, RS.)

What would a parade be without bands? That was not easy to schedule when Ed Trigeiro had the idea of a parade for 1968. The organizing committee contacted a myriad of marching bands without any luck, no response from any group. Fortunately someone learned about the musician's union and their control over all marching units in all parades. Once a check arrived in the union office in New York—to meet the monetary requirements for the full union scale—marching bands became available in Sebastopol. The Apple Blossom Parade has featured marching bands since then. A dapper Analy High School marching band is shown in 1997. (RS.)

Elementary, middle, and high schools in Sebastopol and the surrounding area participate in the Apple Blossom Parade. The bands practice throughout the school year, but marching drill practices are dependent on relatively dry weather and are very important to the success of any parade. A variety of uniforms are seen, from quite casual to formal attire. In spite of what is worn, school pride is uniform. A glance to the top and center of the picture featuring the Apple Blossom Elementary School band shows a couple of parade viewers with special box seats. The Gravenstein Junior High School band with the honor guard and majorettes looks sharp in 1978. (Above, RS; below, WSCHS.)

Parade day arrives. All the months of preparation on the part of many comes together. All city streets crossing Main Street are blocked to thru traffic at 9:30 a.m. No cars are allowed on Main Street. Portable toilets are in place. The reviewing stand is ready, and reserved seats are occupied. At 8:30 a.m., the parade begins staging in the vicinity of Analy High School. Each entry reports to a predetermined location with spaces appropriate to the numbers of participants. Friends are greeted, hairdos touched up, and costumes and uniforms are straightened. Everyone waits as bands tune up, last minute entries look for their location, and rumors fly. More waiting and finally comes the announcement that the parade is about to begin. Then with a blast from the siren of the motorcycle police escort, the Apple Blossom Parade is on its way. The spacing is vital—there must be adequate room allowed between entries so judges can appreciate each as it passes in review. (WSCHS.)

Parents of school-aged children often find themselves involved in many activities. Youth organizations such as 4-H, FFA, Boy Scouts, Cub Scouts, Brownies, Camp Fire Girls, dancing troupes, and baton twirlers involve parents as well as children. It is not uncommon for a family to be represented in two or more parade categories. They must prepare their children for many eventualities. Protection from possible rain, applications of sunscreen, and availability of clean uniforms all must be considered. Those concerns have not changed during the history of the parade, however one hazard is missing. When the railroad tracks were removed from the middle of the street in 1987, there was one less obstacle to threaten marchers and bicyclists. Note the track in the midst of the 1974 Cub Scouts below. (Above, RS; below, WSCHS.)

The 1.5-mile route is a long hike for toddlers and their slightly older siblings. Moms are with them each step of the way, sometimes on the sideline, but always watchful. Sebastopol Mothers, above, is such a group; whether the entrant is in a stroller, on foot, or on a tricycle, mom or dad cannot be too far away. Under threatening skies, this group carries on. The parade makes frequent stops and when moving, does not go very fast. Maintaining balance plus neighbors' balloons present problems for the group below. (Both, RS.)

Floats appear in a variety of sizes and shapes. A lot of time goes into the preparation of elaborate floats that are then powered by a hidden truck or tractor. The horse-drawn entries always have support groups; a back-up crew is needed to keep things tidy. Carriages and wagons are pulled out of barns, dusted off, and given a good polish before being put to use for the day. Riding clubs have a long-standing tradition of appearing near the front of parades. It is obvious that the horses also take great pride in appearing stately. Uniformed riding clubs present their colors below with many riding groups following. (Above, WSCHS; below, RS.)

The "Train Down Main" was the theme of the parade in 1984, paying homage to the city that did have a train that ran right down the middle of Main Street for many years. The finishing touches were put on those tracks in May 1904. On October 10, 1904, the switch at the Sebastopol powerhouse, the current home of Hop Monk Tavern, was thrown, sending power over the wires of the entire Petaluma and Santa Rosa line. From downtown Sebastopol, trains would travel north to the end of the line at Forestville via Molino, Taylor, Graton, Manzana, Ross, and Harbine. At that time, the Northwestern Pacific Railroad owned the steam railroad branch from Santa Rosa to Sebastopol. Southern Pacific diesel locomotive No. 2501 had a place of honor in the 1984 parade, proudly wearing balloons and the banner of the "Apple Blossom Special." The tracks were removed three years later. That reality was faced with much emotion and fanfare. (At right, DM-SCL; below, WSCHS.)

Crowds line both sides of Main Street, and attentive children await the possibility of candies tossed from the floats. Themes are established in advance so that each year's entries can be well planned. "Once Upon a Time," "Aloha," and "No Place Like Home" had their turn as parade themes, as did the category of nursery rhymes. Local service clubs have been consistently supportive of all phases of the apple festival celebration. Private and public schools from the area, church groups, hospitals, and law enforcement join them. Parade participation is only one example of Apple Blossom community involvement. The Sebastopol Rotary and the Interact Club are shown marching below in 1997. (At left, WSCHS; below, RS.)

Sebastopol is a diverse community, and the Apple Blossom Parade is a time for those varied groups to come together. The Sebastopol Newcomers Group is active throughout the year. The parade gives an opportunity for that group, active since 1978, to make its presence known. It is safe to say that the vintage auto is not a newcomer. (RS.)

Palm Drive Hospital's Pink Ladies proudly parade in their rose-colored uniforms. These dedicated ladies are faithful in their attendance as hospital volunteers. The organization began in 1950 under the direction of the Red Cross, first called "Gray Ladies." Palm Drive coats have been upgraded to maroon. Pink Ladies fill many needs in the hospital and are sometimes referred to as the "Giving Heart" of the hospital. (RS.)

There are many choices for float construction. Through the years, groups have attempted to replicate the type of structures similar to those made from apples in the early 1900s. That technique is complicated and time consuming and requires many apples. In 1972, an entry was made to resemble the California State Capitol building (above). Royalty of another era is heralded by medieval costumes, instruments, and banners (below). Elaborate costumes add a flavor of authenticity, right down to the bows on the shoes. A variety of historic themes have prevailed. "The Wild West," "Lights, Camera," and "Train Down Main" were years recognizing historic significance. (Above, WSCHS; below, RS.)

The sound of the siren adds to the drama. There are often contingencies of tractors, farm equipment, trucks, motorcycles, and finally those residents who traditionally follow the parade. There is a thrill to walking down the middle of the street. Sebastopol and neighboring volunteer fire departments find the parade to be a good showcase for new equipment. In 1978, Hessel displays their engine no. 3, and in 1997 Sebastopol honors their fireman of the year. (Above, RS; below, DM-SCL.)

Many family traditions ensure good seats on parade day. The trick is to get up early and be on time to stake out a watching site on Main Street by 6:30 a.m. Then one must reserve spots with blankets and folding chairs, go home for breakfast, and finally return in a leisurely manner. Others are more spontaneous. They wait until all cross traffic is stopped, then, as shown by these experienced parade watchers, set up their camp chairs right across Bodega Avenue for ringside seats. The magic of clowns evokes all sorts of reactions—utter delight, embarrassment, and puzzlement, such as "where did that little dog go?" Year after year, rain or shine, the Apple Blossom Parade represents many things. Each visitor has his or her own expectations and reactions. There is one constant, though: entertaining the crowd. (Both, RS.)

Five

The Ascent and Decline of the Gravenstein Apple

In a *Sebastopol Times* article in 1952, Rudolph E. Oehlmann summed up apple harvest results of years past. Oehlmann started in apples in 1920. The Sebastopol Apple Growers Union had just been reorganized the previous year. The apple industry was now well established since Nathaniel Griffith had planted the first commercial orchard of Gravensteins in the late 1800s.

The Gravensteins were the first apple of the season on the market and there was not enough volume to meet demand. But then, as Oehlmann pointed out, everyone from bankers to merchants planted an orchard. In 1922, the apple market collapsed. Apple production had doubled, but there was a short but sharp depression after World War I, although the development of cold storage kept late apples available into the start of Gravenstein season. More early maturing apples were planted in the Midwest and East. From 1922 to 1942, various efforts were made to stabilize the marketing and production of Gravensteins. Some orchards were abandoned or removed. World War II created a demand again with the lure of high prices for growers. By the end of the war, and with a record crop to sell in 1949, apple growers were again in chaos.

The Early Apple Marketing Agreement tried to regulate the amount of apples brought to the fresh market at any one time with some success. The marketing agreement resulted in growers finally confronting the question of when to harvest Gravenstein apples. In Oehlmann's judgment, "An immature Gravenstein does make a palatable pie or sauce and a welcome change for consumers from held over storage apples. For this use it should be harvested and shipped as soon as it has sufficient maturity to complete the ripening process."

Oehlmann summed up by saying, "Why not face the fact that the Gravenstein is not a dessert apple but is unexcelled in flavor. We should recognize the Gravenstein is primarily a processing apple. The real opportunity for fresh apple sales is in nearby markets with lower preparation and transportation costs and less competition from late apples shipped here from elsewhere."

National Apple Week never caught on in Sebastopol. However, Sebastopol did host the National Apple Institute for an area tour in 1955. Visitors were bused from San Francisco to Sebastopol orchards, historic sites, and apple processors. At that time, California had six distinct apple growing areas. San Diego's 525 acres produced common varieties, and they were consumed in the county. Tulare County grew White Astrachans that came to market in June, ahead of Gravensteins. Their fruit sold mostly in the Los Angeles area. El Dorado County growers, with higher elevation and more rainfall than Sonoma County, had their 307 acres planted in 11 varieties. Mendocino County's Anderson Valley had 819 acres in common and not so common varieties, such as the Gloria Mundi and Winter Banana. They even offered Burbank's Maiden's Blush. Watsonville statistics were not related other than to report that one grower had a self-contained Hallberg-like operation and grew over 200,000 bushels of apples. Sebastopol outproduced them all with 10,276 acres in all apple varieties and one million boxes sent to market. (WSCHS.)

Apple picking in August was a common scene in Sebastopol, but at Luther Burbank's Experiment Farm if any harvesting was done, it was done out of public view. Burbank purchased his 18-acre farm in 1885. He discouraged visitors by charging a fee to any who braved entry. He checked workers' pockets to be sure no plants left without his approval. There were no plant patents granted until after Burbank's death. Burbank helped Nathaniel Griffith identify his newly planted apple trees as the Gravenstein. Then Burbank worked to enhance the Gravenstein with his hybrids, the Winterstein (1910) and the Bonita (1919), also called the Red Maiden's Blush. Burbank's apple varieties were never commercially popular but were used in home gardens. Sebastopol apple processors also experimented with apples with the Barlow Company producing a syrup-like mixture of flavored fruit called Freem, used to make a frozen dessert that could be dispensed from "Softy Frosty" machines. Gravenstein applesauce made up one-third of the fruit base. (WSCHS.)

When Sebastopol held its first apple show in 1910, Luther Burbank was at the height of his career. He became known as the "Plant Wizard" when in 1881, he produced 20,000 prune trees in six months for Warren Dutton. After Burbank died in 1926, Stark Brothers Nursery from Missouri sent Joseph Keil to be the caretaker at the Burbank Experiment Farm and retrieve as many usable plants as possible. In the 1970s, Stark Brothers was still marketing eight Burbank fruits, among them the Santa Rosa plum, the Van Deman quince, and the July Elberta peach. The Luther Burbank trademark was kept registered and Elizabeth Burbank was paid royalties each year for the use of the name. (Both, WSCHS.)

Gene Bologna, Sonoma County sealer of weights and measures in 1962, weighs an oversized apple while the Hallberg daughters, Donna (left) and Kathy, look on. The Weights and Measures Department is part of Sonoma County's Agriculture Commission that have kept apple and other farm product statistics from 1919 to the present. Previously, apple crop statistics were duly reported in the *Sebastopol Times*, as apples became the most important crop in the area. In 1911, the Sebastopol Apple Growers Union sold 28,597 boxes of apples. By 1936, they were selling 7,667,673 boxes. Twenty-five railroad cars left Sebastopol each day filled with fresh, canned, and dried apples. Sixty dryers were in operation, and $300,000 in wages was disbursed into the local economy. In the 1960s, there were 5,094 acres of Gravensteins producing 49,409 tons valued at $2,734,496. Late varieties of apples were grown on 3,255 acres. But grape acreage was catching up at 10,705 acres with a crop worth $2,113,020. (DHM.)

In 1982, apple growers' wives formed the Apple Publicity Promotion Ladies Effort Corps (APPLE). They enlisted the help of the Apple Blossom Queen, Juli Pimentel. Pictured here at the Hurst Twin Hills Ranch, she was requested to visit fairs, exhibitions, and conventions with apples in hand, be they fresh, juice, dried, or sauce. One event the APPLE Corps hosted was a "Day in the Country" for the media. The group met to view the Dutton's packing operation, visited the cooperative cannery, walked through the Bests' orchard, and then picnicked at the Chenoweth Ranch. The roasted pig on the menu even featured a Gravenstein apple in its mouth. Apple Blossom queens were chosen from seniors in the local high schools. Their duties were minimal but focused on promoting the Gravenstein apple. However, Pimentel also represented California at the annual Miss Apple Queen USA contest held in Gettysburg, Pennsylvania. (DM-SCL.)

Emily Pinnow revels in the 1992 crop of apples. In 1951, teen Joan Voit Chilton poses in her family's apple orchard in the Cunningham area. The Voit Ranch was located off Lone Pine Road. The Voits were one of hundreds of small apple farmers whose apple orchards still surrounded Sebastopol in the 1950s. Emily would have been too young to work in any of the packinghouses in Sebastopol in the 1950s, but Joan might have been eligible. In 1953, an adult education class was held at Analy High School to train fruit packers and handlers. Transient labor could no longer be counted on, and the packinghouses hoped to train new local adult workers and teens to fill their labor needs. (At right, CC; below, JVC.)

In 1920, a Petaluma Chamber of Commerce delegation visited Sebastopol in a special excursion Petaluma and Santa Rosa Railroad car. They arrived at the Sebastopol Apple Growers Union packinghouse for a tour. They were probably not planning on joining the several thousand men, women, and children who were bringing in the harvest in the local apple orchards. Old-timers recalled the aroma of apples everywhere in Sebastopol during the August harvest season. A concentrated apple aroma certainly surrounded the many Sebastopol apple dryers. In 1917, in one day alone, 150 tons of dried apples were sold. Fruit brokers George McFarlane and James Kelly alone bought 145 tons. The per-pound price was 14¢, the highest price received so far. In 1920, there were 6,100 apple-bearing acres but another 11,050 acres had been planted. One acre of orchard with 50 trees could produce 2 to 13 tons of apples. As reported in 1892, dozens of young orchards had been planted in the previous year and were growing all around Sebastopol. (SCL.)

Sebastopol apple growers have always been keen to promote their product to the world. A delegation from the Sebastopol Chamber of Commerce and the Apple Blossom Festival Committee visited with Gov. Ronald Reagan (left) in 1968 to share samples of Gravenstein apples. Joining in the visit is state senator Randolph Collier (far right). The Panama Pacific Exposition in San Francisco in 1915 provided an earlier promotional event with Luther Burbank participating at Sebastopol Apple Day. Burbank waved a wand at a gigantic apple, it opened, and out stepped Elva Howell, the Gravenstein Apple Queen. Sebastopol held its own apple fair that year, too. Burbank was not above selling his product to the public as well as promoting the Gravenstein. (SCL.)

U.S. Congressman and Sebastopol native son Hubert B. Scudder (1949–1957) (shown second from left) enjoys Gravenstein applesauce with Speaker of the House Joseph Martin (far right), and colleagues. The photograph is autographed to Oscar Hallberg. Scudder was born in Sebastopol in 1888. He attended local schools. His business was insurance and real estate. Scudder served as Sebastopol city councilman, and in 1926 held three offices simultaneously as mayor, fire chief, and a seat in the California State Assembly. Scudder helped revive the Gravenstein Apple Show in the 1920s and served as apple show president from 1928 to 1932. Civil War veteran G. A. Blank visited the 1928 apple show and was poetically inspired: "The day was fine, the weather cool, decked with bunting, gay Sebastopol. Let's not forget that merry-go-round. Astride of it the Blanks were found. It was no wonder, at five cents a drink, delicious cider, Volstead act got a kink. I straddled a horse, Mrs. B on a cow. Oh! Gee! We certainly had one bow wow." (WSCHS.)

Jone Ann Pedersen, Miss California in 1949, cuts the ribbon on the Sebastopol Apple Growers Union first railroad car of Gravenstein apples shipped that year. Perhaps pretty girls in bathing suits were needed in 1949 to promote apple sales, a record crop year. The dried fruit market, usually 40 percent of the crop, was depressed in those postwar years. Sebastopol's 800 growers expected to have a one million box surplus. They were facing a multi-million-dollar loss. Growers applied direct marketing tactics by hauling fruit by truck to the city themselves. They had choices to handle their freight. Before the Petaluma and Santa Rosa Railroad arrived in Sebastopol in 1904, growers had to haul fruit by wagon to Petaluma where it was loaded onto ferryboats bound for San Francisco. The railroad opened up more possibilities for marketing the crop. In 1914, Sebastopol shipped 250 railroad cars of apple products to the eastern United States with 300 railcars leaving the area to be sold in California alone. (DRC.)

In 1949, a group of 23 growers formed the Sebastopol Cooperative Cannery to process their excess apple crop. They invested in a modest amount of equipment and canned 135,000 cases of applesauce. They employed 300 people at peak season with an annual payroll of $250,000. Their gross income was $2 million. In 1954, the cooperative built a $0.5 million cannery that could produce 1.5 million cases of applesauce. They then added juice and sliced apple products to their line with all operations being mechanical. Cooperative membership grew to 70. Many members also belonged to the Sebastopol Apple Growers Union. This cannery was the first to negotiate a settlement with the Teamsters in 1956, a union inroad that helped break the strike. Other canneries followed their lead and signed with the union. The cannery went bankrupt in 1983. (SCL.)

What do you do when your apple trees produce more fruit than you normally have an outlet for? When that happened in 1949, Sebastopol apple growers went all out to enlist San Francisco's help in using and enjoying the Gravenstein apple. Sebastopol high school girls rode on top of a caravan of trucks filled with Gravensteins and threw apples to passersby near the Ferry Building. Among the helpers were, from left to right, Joan Hotle, Lenore Helwig, Rae Eusitalo, Barbara Starr, and Peggy Vice. San Francisco department stores, bakeries, and restaurants all pitched in by featuring the Gravenstein for sale or use during those 10 days in August. Companies purchased apples for employee snacking, and San Francisco newspapers helped by printing recipes. Home cooks were urged to use Gravensteins everyday as pie, cobbler, baked, or just eaten fresh. Even First Lady Elizabeth Truman supplied the recipe for the president's favorite apple concoction. San Francisco's city health director, Dr. J. C. Geiger, added his medical advice too, "Apples should be a part of everyone's diet." (SFEX.)

It was a sad day indeed in 1999 when the very visible, historic orchards of the Hallberg family were bulldozed to make way for a vineyard. That same year, Vacu-Dry decided to close up operations as an apple processor. Vacu-Dry had been in business for more than 60 years. More than 25 percent of Sonoma County's total apple tonnage of 40,000 could not find a home. According to the Agriculture Crop Report of 1998, grape acreage netted more income than apples even though orchards produced more tonnage. Vacu-Dry was the second-largest dried fruit processor in the United States. They also dried other fruit, which was used in cereals, muffins, soups, and frozen entrees. Vacu-Dry sold its equipment and inventory to Tree Top, Inc., of Selah, Washington, for $12 million and kept their two plant locations to rent out to other businesses. (EM.)

Fruit crate labels acted as publicity for the product inside, but today many see them as collectible art. Thomas Patrick Jacobsen in a 1986 *Sebastopol Times* article discussed the fruit crate label and its place in California agricultural history. He said 1,872 different labels were lithographed in San Francisco by more than 50 printers. From the 1880s to 1920, labels had a naturalistic and historical theme. In the 1920s to 1930s, the health benefits of the products were emphasized. Then in the 1950s, labels became more artistic. After 1950, fruit was packed in preprinted cardboard cartons and unused labels sat at printers or in packing sheds. Eventually collectors discovered the labels. Sebastopol processors used 24 different brand labels for various apple products. The Furusho Brothers used the Dickie Girl and Jackie Boy labels, and the Redwood Empire label denoted O. A. Hallberg products. The Sebastopol Cooperative Cannery, representing 34 growers, used the Appletime, Apple City, and Apple Land labels. (WSCHS.)

The years of World War II provided a boom time for Sebastopol apple growers. The U.S. Government purchased apple products for military use. Even during the Vietnam War, local Vietnam veteran Steve Stedman was delighted to find a touch of home when served Hallberg apple juice. But in postwar years, growers needed to find new markets or scale back production. The Sebastopol Apple Growers Union scaled back with the dismantling of its packing operation to make way for the Sebastopol Veterans Memorial Building in 1956. The Sebastopol Apple Growers Union still had its other location on Occidental Road. The apple crop's value was beyond anyone's expectations in those war years. The income from all apple products, including fresh, canned sauce, vinegar, brandy, and other by-products were as follows: in 1942, $1,697,935; in 1943, $5,070,550; in 1944, $4,553,020; in 1945, $9,288,096; and in 1946, $5,974,562. (SCL.)

The plight of the Gravenstein apple in Sonoma County has been taken on by the local Slow Food chapter. The Gravenstein apple project has been designated a *Presidium* (Latin for "fortress") by the Slow Food Foundation for Biodiversity, based in Italy. The Russian River Slow Food chapter seeks to promote the consumption of Gravensteins at restaurants and farmers markets. There are less than a dozen commercial Gravenstein growers in the Sebastopol area now and one processor, Manzana, in Graton. There were 875 acres of Gravenstein apples being grown as of 2008. Some of the remaining growers have gone organic as that market appears more viable. Sixty percent of all the U.S. apple crop is sold fresh at prices three times that paid for processed apples. Sonoma County has only sold about seven percent of its apples fresh. Foreign competition has cut into the processed markets with devastating results for Sebastopol's growers. More information about Slow Food activities can be found at www.slowfoodrr.org. (SFRR.)

MAP SEARCH